Thomas Wardle

On the Entomology and Uses of Silk

With a List of the Families, Genera, and Species of Silk Producers Known

up to the Present Date

Thomas Wardle

On the Entomology and Uses of Silk
With a List of the Families, Genera, and Species of Silk Producers Known up to the Present Date

ISBN/EAN: 9783337222055

Printed in Europe, USA, Canada, Australia, Japan

Cover: Foto ©berggeist007 / pixelio.de

More available books at **www.hansebooks.com**

NORTH STAFFORDSHIRE NATURALISTS' FIELD CLUB
AND ARCHÆOLOGICAL SOCIETY.

ANNIVERSARY ADDRESS

OF THE PRESIDENT,

THOMAS WARDLE,

FELLOW OF THE CHEMICAL SOCIETY; FELLOW OF THE GEOLOGICAL SOCIETY:
FELLOW OF THE IMPERIAL INSTITUTE;
MEMBER OF COUNCIL OF THE PALÆONTOGRAPHICAL SOCIETY:
CHEVALIER DE LA LEGION D'HONNEUR OF FRANCE;
OFFICIER D'ACADEMIE OF FRANCE;
MEMBRE DU JURY DE L'INDUSTRIE DE LA SOIE À L'EXPOSITIONS
UNIVERSELLES À PARIS, 1878 AND 1889;
HONORARY SUPERINTENDENT OF THE INDIAN SILK CULTURE COURT OF THE
COLONIAL AND INDIAN EXHIBITION, LONDON, 1886;
CHAIRMAN OF THE SILK SECTION OF THE ROYAL JUBILEE EXHIBITION,
MANCHESTER, 1887;
ONE OF THE EXAMINERS TO THE CITY AND GUILDS OF LONDON INSTITUTE;
PRESIDENT OF THE SILK ASSOCIATION OF GREAT BRITAIN
AND IRELAND;
PRESIDENT OF THE NORTH STAFFORDSHIRE NATURALISTS' FIELD CLUB
AND ARCHÆOLOGICAL SOCIETY.

ON THE

ENTOMOLOGY AND USES OF SILK;

WITH A

LIST OF THE FAMILIES, GENERA, AND SPECIES OF

SILK PRODUCERS,

KNOWN UP TO THE PRESENT DATE.

Newcastle-under-Lyme:
G. T. BAGGULEY, PRINTER, HIGH STREET.

THE ENTOMOLOGY AND USES OF SILK,

BY

THOMAS WARDLE, F.C.S., F.G.S.

I feel that a great compliment has been paid to me to-night in having been unanimously asked to occupy the Presidential Chair for another year.

I have accepted this honour with some reluctance, because it is possible for one man to be in the same office too long, and that more useful members are thereby prevented having the opportunity of giving valuable services to this Association, and through it, to Science.

I beg, however, to thank you all, ladies and gentlemen, for your confidence in me, which is, at the same time an endorsement of your satisfaction for the past year, and your trust in me that I shall do my best during the present year.

My affections are very much bound up in this North Staffordshire Field Club and Archæological Association, for several important reasons.

First, that we are all lovers of Natural History, and admirers of the glorious world of nature ; collaborateurs of those in the past who have tried to discover some of her mysterious laws, and those of the present and future, who are, and who ever will be, finding for themselves untold pleasure in the investigation of those forces which have such manifold play, and with such varying and wonderful results as can only be completely known to the Great Architect of a Universe who made " all things visible and invisible ;"

from the worlds which we are permitted to see afar, to those beyond, in illimitable space, to the lesser world in which we live, and which we are permitted to examine, and to take pleasure therein.

I have often thought of that unique and comprehensive expression in the creed of St. Athanasius, "maker of all things visible and invisible." Of things visible, how much more do we not now know than the framer of the phrase, of the constitution of the earth's crust; the structure, functions and properties of plants, the whole animal world, and the insight we have been able to obtain as to their composition by the aid of chemical science.

Of the " things invisible " what vast revelations have been given to us by the microscope, since the day which first heard the recitation of that article of our faith, revelations even to the very gate of that knowledge by which the origin and cause of life is centred.

The revealing to sight of that formidable world of bacteria, their functions, uses and dangers, is progress which would hardly be credible to the old observers of natural science and history. Our love for this progressive knowledge is inborn; from Adam, who is recorded to have had the beasts of the field and the fowls of the air brought to him to see what he would call them; "and the man gave names to all cattle, and to the fowls of the air; and whatsoever the man called them, that was the name thereof." How, since that ancient past, has that naming been continued; and how we are yet infected with the desire to find a new species, and to add a new name to an already vast nomenclature!

Of Solomon it is written that "he spake of trees, from the cedar tree that is in Lebanon even unto the hyssop that springeth out of the wall. He spake also of beasts, of fowl, of creeping things, and of fishes." (1st Kings, iv, 33 v.) And we are reminded, at least those of us who heard of our friend Mr. Carr's eloquent recital of the Talmud story of the bees, on the Sunday afternoon under the trees at Clumber last year, in words partly his own, of Solomon's knowledge of insects.

As our Annual Meeting is not rigidly confined to Science, but a friendly occasion to report progress, perhaps you will pardon me if I read these verses; they are sure to interest at least those who have not heard them before:—

5

When Solomon was reigning in his glory,
Unto his throne the Queen of Sheba came,
(So in the Talmud you may read the story)
Drawn by the magic of the monarch's fame,
To see the splendours of his court; and bring
Some fitting tribute to the mighty King.

Nor this alone ; much had her Highness heard
What flowers of learning graced the royal speech ;
What gems of wisdom dropped with every word ;
What wholesome lessons he was wont to teach
In pleasing proverbs ; and she wished, in sooth,
To know if rumour spoke the simple truth.

Besides, the Queen had heard (which piqued her most)
How through the deepest riddles he could spy ;
How all the curious arts that women boast
Were quite transparent to his piercing eye,
And so the Queen had come—a royal guest—
To put the sage's cunning to the test.

And straight she held before the monarch's view,
In either hand, a radiant wreath of flowers ;
The one bedecked with every charming hue,
Was newly culled from nature's choicest bowers ;
The other, no less fair in every part,
Was only wrought by imitative art.

" Which is the true and which the false," she said.
Great Solomon was silent.   All-amazed,
Each wondering courtier shook his puzzled head,
While at the garlands long the monarch gazed,
As one who sees a miracle, and fain,
For very rapture, ne'er would speak again.

'· Which is the true," once more the woman asked,
(Pleased at the fond amazement of the king,)
" So wise a head should not be hardly taxed,
Most learned liege, with such a trivial thing ; "
But still the sage was silent ; it was plain
A deepening doubt perplexed the royal brain.

While thus he ponders, presently he sees,
Hard by the casement (so the story goes),
A little band of busy, bustling bees
Hunting for honey in a Sharon rose.
The monarch smiled, and raised his royal head ;
"Open the window !" that was all he said.

The window opened at the King's command,
Within the room the eager insects flew,
And sought the flowers in Sheba's dexter hand.
And so the King and all the courtiers knew
That wreath was natures ; and the baffled Queen
Returned to tell the wonders she had seen.

6

My story teaches (every tale should bear
A fitting moral) that the wise may find
In trifles, light as atoms in the air,
Some useful lesson to enrich the mind ;
Some truth designed to profit or to please,
As Israel's King learned wisdom from the bees !

Apropos of bees and apiculture, I learned the other day from the Board of Trade, that a bee must visit 218,750 flowers for each ounce of honey gathered, and that the largest beekeeper in the world is Mr. Harbicon, of California, who possesses 6,000 hives, supplying 200,000 lbs. of honey per annum. The United States is the greatest honey producing country in the world ; it has 2,800,000 hives belonging to 70,000 rearers, and producing annually 61,000,000 lbs. of honey. An enemy of the bees is a friend of mine, the larvæ of the moth Galleria cerella, creeping into the hive and boring into the honey cells, feed upon the honey ; they make silken cocoons in the cells from which the moths emerge, as you may see in this case.

In the two-fold objects for which the Field-Naturalist part of this Association exists, namely, the praise of nature by our observation of it in our excursions, and in the examination into its constituent parts, we have the encouragement of a greater teacher than the two I have just mentioned ; He who bid us consider the lilies of the field ; how they grow ; that they toil not, neither do they spin, and that yet Solomon in all his glory was not arrayed like one of these ; and in so considering do we not forget our toil and the care of to-morrow whilst enjoying the collective and diversified enquiries which occupy us in our monthly Excursions, both amongst the lilies of the field and the birds and denizens of the air and landscape ; but our attention and pleasure is not confined to the history which is termed natural ; there is also a living history, so to speak, of our own race as seen in the various forms of sepulture, defense, worship, and of later times, of residence, with which our County abounds.

To these, not only our interest but our study and care is given, and the County owes us no small debt of gratitude for bringing before it, by patient and long continued investigation and description, the instruction to be derived and imparted by those of our members who so very ably and lovingly give themselves to the duty.

These are our raisons d'être for the establishment and continuation of this Association, whose work last year will I think compare favourably with that of the past, and whose future is full of promise and encouragement to both young and old enquirers after truth.

Another thought should give us hope and determination. The example of the untiring and valuable work done in our Association, by the late Mr. Garner especially, and many others whose record is to be found in our annual reports. Amongst others of our present time I shall mention only four out of a goodly company of zealous workers :—Drs. Arlidge and McAldowie, our former Presidents, who have so distinguished themselves by their thoughtful and learned contributions to our Annals ; Mr. Lynam, whose portrait ought to be painted with a drawn sword in his hand, keeping off the restoring vandals from our ancient camps and beautful mediæval architecture, all traces of which he so jealously guards, in order that we may enjoy what I call the visible and veritable history of our County unmarred by the falsifying process so happily expressed by the Germans "Restaurirt ist ruinirt."

Lastly, no greater encouragement can be given to the continuance of our collective work than in the splendid constancy and devotion given for nearly thirty years by our excellent Secretary, the main-spring of our success and work. He has the happy knack of welding together all sorts and conditions of men and women into a joyous and harmonious whole, until we know nothing of "classes and masses," and I am sure I shall be expressing the opinions of every member of our Club and Association, that his name will live in the annals of the County long after many of its principal actors are forgotten. To him our earnest and grateful thanks are due, and I have long wished for an unbidden opportunity of expressing my convictions in this respect.

With such associationship as I have mentioned, permit me to advocate a devoted and earnest determination of each and every member, particularly the young, to try to be present at each Excursion, and if possible to beat past records in field-work and observation.

There is much to do that is new ; our sectional work needs strengthening. I think the Geological Section, if I may say so, carried off the palm last year. I wish the members of other Sections could catch the enthusiasm which inspired us last year in tracking a volcanic dyke mostly hidden, from Trentham to Swinnerton, where we may say we fell literally and really into a trap, well a trap-pit, in the shape of a huge quarry of upheaved larva, and returned laden with ashes, and information healthily sought.

Having said this much, mainly hoping to stimulate our new and younger members to follow up the work of the older votaries of knowledge seeking, I will turn to the subject of my Presidential Address.

I have selected for consideration the study of a delightful part of entomology, which has a beautiful, important and economic outcome. The title of my paper is "The Entomology and uses of Silk."

I have selected this subject because it is one at which I have of late worked most, and in both aspects I have for some years been collecting silk-producers, not so much for their interesting larval forms, or for the great beauty and variety of their imago states, but in order to study the structure and physical properties of the fibres which their larvæ seriposit for a covering and protection during the time they assume their torpid chrysalide state, a covering, which, to serve economic ends, is from a few species, ruthlessly torn by man to provide in other forms and in other colours, fabrics and trimmings for human coverings and adornment, as well as of thread to sew them with.

When I began this delightful and interesting study many years ago, my knowledge of silkworms, their moths and cocoons, was limited to about 20 species, and they chiefly Indian. It was then known only to a few entomologists that there existed many species of silk secreting catterpillars in England and elsewhere. To-day, after a world-wide search, I am able to hand a new list to our Field Club of 27 families, numbering 170 genera, with at least 630 species, all of which make silken coverings, but often as varied in shape, size, build and differences of fibre, as the appearance of their moths, the last stage in the curious life-history of this class of insects.

Of these 630 species, there are some exhibiting wonderful characteristics. There is the tiny caterpillar of the smaller moths, not more than half an inch in length, and from this can be shown larvæ of all sizes, up to the giant Tussur, two of which I took out of a Terminalia Tree in a Missionary's garden in the Manbhoom Jungles of Bengal, each 5 inches in length and $\frac{3}{4}$ of an inch in thickness. I hope to show a representation of one of them on the screen presently. Then there are the catterpillars which lead a solitary life, and shroud themselves in their silken envelopes when they happen to have arrived at caterpillar maturity; not so with other species, for there is the Cricula Trifenestrata of India, which lives in companionship with its fellows, and when cocoon-making time comes, they work side by side, interlacing their silken threads until they lie hidden in a colony as you see them here.

Others like this specimen from Africa, first weave a bag of silk, then go inside to envelop themselves singly with a separate cocoon. This bag is full of them.

Another species is of a more military order. Its members never take their walks alone or in pairs, but always in line, one leading the rest, even if hundreds are following and so closely that they look like a long length of coloured cord along the ground or upon a tree, a most remarkable sight of gregarious habit.

I brought some home last year from the woods of Monte Carlo, and one evening I went to my greenhouse and found them on their slow march exactly in a straight line, in single file, nose to tail, along the top of the hot water pipes. In the Riviera you may see hundreds of their bulbous cobwebby bunches in the fir trees ; these are their hiding places. They rest in them in the day and come out at night to feed upon the firs. These white bunches are their houses of silk, tied from all points to the fir branches Some of them are larger than footballs. The insect is known by the name of Bombyx, or more properly, Cnethocampa processionea. Its silk has not been turned to any profitable account, but it could be used for spinning ; it could not be reeled like the cocoon of the Bombyx Mori.

There is a caterpillar in Demerara whose legs only go half the length of his body ; he goes about like a snail with his house on his hinder parts, the leg portion of his body for crawling, the legless part carrying his cocoon.

But I should weary you if I attempted to recount a hundredth part of the list up to date I have brought here.

With the multitudinous variety of moths you are well acquainted, with their enchanting shapes and colours, and with their pretty patterned wings, so I will content myself with showing you a few of the more important and interesting species on the screen at the conclusion of my paper.

Silk is also produced by many species of spiders, also by the larvæ of the Ichneumem flies. On the table is an interesting collection of these flies whose larvæ have laid their eggs in the bodies of caterpillars. After the caterpillars have made their cocoons and passed into the chrysalide state, the tiny ichneumen eggs hatch and their larvæ surround themselves with a silken thread 6 or 8 times finer than the Bombyx fibre, until a small cocoon results in considerable numbers not larger than rice grains, and occupying the interior of the large cocoon to the destruction of the chrysalis. ⁣ ¡In day time these ichneumen flies emerge from their little silken shrouds, and eat their way through the silken prison of the larger cocoon.

Another class of insects, the Coccidae much studied by Mr. Newstead of Chester, also make tiny cocoons which very much resemble silk, under the microscope, but the fibres, which dissolve in turpentine, prove them to be of a waxy and not a silken nature. There is a small collection on the table. These Coccids are found on grass and on the bark of trees.

And now I will as briefly as I can, touch upon their uses to man. First comes the little Bombyx Mori, whose caterpillars yield us the ordinary silk of commerce, from which the finest fabrics are made ; whose ancestors found such favour with the wives and daughters of Chinese Emperors thousands of years ago, that they delighted themselves in unwinding the silk from the cocoons and in making it into threads for embroidery and for textile work, also in dyeing it, and afterwards in weaving it into patterned fabrics. For centuries in ancient times, it was the pleasure and pride of high-born ladies to broider and weave. Thus in the 3rd Illiad :—

> " Meantime to beauteous Helen from the skies,
> The various goddess of the rainbow flies.
> Here in the palace at her loom she found,
> The golden web her own sad story crown'd ;
> The Trojan wars she weav'd, herself the prize,
> And the dire triumph of her fatal eyes."

I have here two very small pieces of silk, both found in Egypt in ancient Christian Coptic graves by my friend Mr. Flinders Petric, the renowed Egyptian explorer. One is of the 4th century, found at Achmim, and is ornamented all over with Maltese-shaped crosses. I have had it reproduced in Germany. The other is of the 6th century, found at Fayum ; it is a very interesting specimen of weaving in stripes It is, too, remarkable for the beauty and permanence of the dyes to which it owes its colour. I have brought a reproduction of it, recently woven in Scotland, the herring-bone texture of weaving is copied, also the stripes and the dyes and colours are exactly reproduced. That the colours should have stood unfaded for thirteen hundred years is marvellous, and proves the excellence of their dyes, and the high perfection in tinctorial art of their dyers.

SCRAPS OF SILK HISTORY.

More is known of the early history of silk in the West than in the East, but dating only from Classic Greek and Roman times.

Aristotle wrote on the life changes of the Silkworm, and gives to Pamphile, the daughter of Plates, the honour of first reeling silk on bobbins for weaving into gauzy fabrics, which made the Island of Cos famous in history and song, for the Coan vest was the most prized garment of the classic lady. It was so transparent that it allowed the body to be seen through its clinging folds; and Horace says of it :—

> "As if unclothed she stands confessed
> In a transparent Coan vest.

Tibullus speaks of a Coan vest for girls :—

> " She may these garments wear, which female Coan hands
> Have woven, and in stripes disposed the golden bands."

A statue to Vertumnus had this inscription on its base :—

> " My nature suits each changing form,
> Turned into what you please I'm fair,
> Clothe me in Coan, I'm a decent lass,
> Put on a toga, for a man I pass."

Plutarch dissuades the prudent wife from wearing silk. Martial speaks of silken fillets for the hair and other silken goods being sold at the Viscus Tuscus at Rome. Galen recommends silk thread for the tying up of blood vessels ; America now parades this as her new surgical invention ; verily there is nothing new under the sun ! Heliogabalus was the first Roman to wear a holosericum or robe of silk ; he also kept by him a silken rope of purple and scarlet to hang himself with when the day of his destiny was over. In the reign of Tiberius, women of rank only were allowed to wear eastern silks, "oriental sericum." Aurelian refused his Empress a silken shawl she coveted, because of its costliness, weight for weight in gold. But Julius Cæsar used silken curtains, and wore silk, when he appeared in public, dyed with the purple of Tyre, the product of Murex Trunculus and Murex Brandaris, two species of Mediterranean shell-fish. There is also a third, Purpura lapilla, and all are found from Italy to Tyre. A Nicaraguan species, named **Purpura patula**, and our own common whelk, also yield the dye. In the days of Marcus Aurelius Antonius, shawls and robes, and scarves had so accumulated in the silk presses of the various empresses, that he, being a philosopher and economic ruler, sold them all by public auction in the Forum of Trojan, to help the exhausted treasuries. The early Christians protested against the luxury of silk. Cyprian, the Bishop of Carthage, said, "Those who put on silk and purple cannot put on Christ."

There is preserved in the Edict of Diocletian in A.D. 303, the maximum prices allowed by him for tailoring in silk, as follows:—

To the tailor for silk lining a fine vest... ..... 6 denarii.
To the same for an opening and edging of silk 50 denarii.
To the same for an opening and edging with
  stuff made of a mixed tissue of silk and
  flax .... 30 denarii.

Mrs. Lynn Linton, the accomplished novelist and authoress, who has written three charming and learned papers on silk in the "Queen" for the month of January, which I cordially recommend you all to read, and from which I have gathered the above historical particulars, thus winds up her first paper :—" Silk, then, was both known and used in the later days of Rome. It was the sign of corruption and effeminacy. It had to fight its way into general acceptance against all the forces of simplicity and virtue which were associated with wool and linen arrayed against silk. But the wheel rolled on, and the prohibitions and prejudices were finally removed, till now the very beggar-woman at your door has something of silk about her, and the material which an emperor refused to his empress, the vagrant and the pauper toss on to the dust-heap when they have done with it."

Coming down to mediæval times, we find Constantinople the chief seat of the European silk trade. In Justinian's time it had got no further westward. He limited the price of silk to £4 15s. 9d. per lb., and when dyed imperial purple its price was quadrupled. The Christian clergy gradually became captivated by its beauty, and blessing superseded anathema. Abbot Benedict, in 685, went to Rome to buy silks. St. Cuthbert, we know, was buried in the 7th century at Landisfarne Island, and was in the 13th century disinterred and re-buried at Durham Cathedral, first being re-robed in silks which may be seen now in that Cathedral, and reproductions of which, by the kindness of Canon Greenwell, I have been able to print and bring here. And so on up to to-day, the Church has become the nursing mother of all that is beautiful in design, and colouring, and in texture of silk, and so it should be to give God our best.

Let us turn now to the silk of the present day. I will not weary you with the history of the introduction of silk manufacture into England. I have here a specimen of blue and red silk from a deed of the time of Richard I., A.D. 1190, which was used as an attach-ment to a large wax seal. Both the silk and the dyes are interesting. The blue is woad, the ancient English blue dye. The poet Dyer, who wrote in the early part of the last century, thus mentions it :—

> " Our valleys yield not, or but sparing yield
> The dyers' gay materials. Only weld,
> Or root of madder, here, or purple woad,
> By which our naked ancestors obscur'd
> Their hardy limbs, inwrought with mystic forms
> Like Egypt's obelisks."

The red is Kermes, an ancient colour produced by the oak-feeding insect, Coccus quercus. It was generally used for this purpose in those days in Europe before the introduction of Cochineal from Mexico into Spain by the Spaniards in 1543, and practised at Bow, where it became celebrated as the Bow dye. Its intro-uedtion into Italy was not until 1548, five years later. It was soon the downfall of Kermes, which, although generally now disused, is still found and used in Tripoliza in Greece. Pardon this digression, but it is difficult for a dyer to disassociate dyeing from silk. From the time when the persecuted Huguenots brought to us them-selves and their trades at the time of the persecution of the protestants in Flanders and France in the latter half of the sixteenth century, and again in 1685, at the revocation of the Edict of Nantes, we have been a silk manufacturing country. These epochs gave us a silk industry, as well as many other arts and crafts. The old rhyme sings :—

> " Hops, Reformation, boys, and beer
> Came into England all in a year."

Very varying have been the vicissitudes of our silk industry. Spitalfields, almost the oldest centre, still maintains its artistic prestige as these gorgeous and artistic surroundings show. My friends, Messrs. Warner and Sons, have to-day done the Field Club a signal service in lending these beautiful examples of brocades, damasks and brocatelles, unrivalled for texture, pattern and colouring by any silks of the kind I have ever seen in France, and my opportunities of judging, having served on the Silk Juries of the great Paris Exhibitions of 1878 and 1889, have not been small. The same must be said for these splendid brocades and figured stuffs for Church furniture and dress purposes, made and kindly lent to us by my artistic friend Mr. George Bermingham, Silk Manufacturer, of Leek. These will show that England can and does manufacture silk equal for every possible requirement, and emphatically bears protest against the false assertions of the middleman that good silks can only be had from France.

One of my objects in choosing this subject, has been the hope of enlisting your sympathy and patriotism, the ladies particularly, in favour of British silks, and to ask you to help the efforts of the Silk Association of Great Britain and Ireland (of which I have the honour of being President), and of the Ladies National Silk Association, presided over by H.R.H. the Duchess of Teck, with Lady Egerton of Tatton, as Honorary Secretary, whose object is to solicit the membership of all ladies who will agree only to ask for British silks when they want to buy silk of any kind; still not pledging themselves to buy such silk if they do not find it as cheap, as durable, and as artistic, as foreign silk.

This Ladies Association is spreading rapidly; already the looms of Spitalfields and Macclesfield are again busy, and we hope for a great revival in the other time-honoured centres. The looms of Spitalfields and Patricroft were merry with the sound of shuttles only a few sad months ago, when the Duchess of Teck gave orders that all silks for the Royal wedding should be home made and not foreign. Here is a specimen of it with May blossoms woven in silver. Alas! that the looms should have been stopped in the middle of their work and their shuttles made to weave for woe instead of bridal joy!

In conclusion, I am reminded of the Ancient Grecian story of Minerva and Arachne, which tells that in the days when purple was rare and highly appraised; the dye very hard withal to extract from its tiny gland in the shelly Murex, which but hardly yielded its tinctured drop of costly colour, there lived by Lydia's shores, Idmon the purple dyer, in his craft renowned for skill. He had a daughter whose name was Arachne, famed for her art in spinning and broidery, but who acknowledged not her obligations to and dependency upon, great Jove's daughter, Minerva; (in the Greek mythology Minerva was called Pallas Athene, the goddess of the liberal arts); but disdainfully challenged her to a contention of skill. They met, the story tells, to weave their separate thoughts with fixed warp and bobbins of weft, each in her own loom. It was a pleasure to observe Arachne winding the wool and curling and twisting it into fine threads, ready to contend with Minerva in the trial of her skill in weaving. Each put her loom in a separate place and stretched the fine threads thereon. The combatants hastened to their work having first girded their robes about their bosoms. They ply their deft hands, and their facile brains encourage the work. Minerva weaves the Castle of the Cecrops standing on the rock of Mars and of the quarrel concerning the name of the country. Twelve immortals are seated on their thrones in austere solemnity with Jupiter in their midst. The individuality of each god is delineated and Jupiter is rendered in regal splendour. Neptune the sea god alone is standing, and with his trident he strikes the unhewn rock from which the saltwater gushes forth. Minerva is shown furnished with the defending Ægis, and having on her head a helmet and in her hand a pointed lance. At the place where the lance has pierced the ground a green olive tree bearing berries is sprouting. The work is surrounded by a garland. The gods look at it with astonishment.

But Arachne wove the story of Europa carried away by the bull. The latter seems to be really living and the sea to be heaving. In addition Arachne wove Asteria seized by the flying eagle, the loves of Leda and the Swan, and of Antiope and the false Satyr. She wove the scenes of Jupiter as Amphitrion wooing Alkmene ; Mnemosin's seduction by Jupiter in the guise of a herdsman, Aegina and Jupiter in the fire ; and lastly Danæ tempted by gold and Proserpine by a Dragon. An ivy garland went round the border with flowers interwoven. The nymphs said about the woven work of art that neither Minerva nor jealous mind could find fault with the work ; Arachne's fertile skill brought her to Minerva's level, but the latter's work was inspired by a nobler mind. Notwithstanding Dante meeting her in Purgatory says :

> " Although in colours variegated more,
> Nor Turks nor Tartars e'er on cloth of State,
> With interchangeable embroidery wove,
> Nor spread Arachne o'er her curious loom."

Minerva viewed Arachne's work, and incensed, rent it into pieces, chastising its founder with her broiding bobbins, whereat Arachne, so keenly feeling her degradation, desired not to outlive it, took a rope to a neighbouring tree, and hasted to end her troubled life. As she in mid-air hung, Minerva with her wand changed her into a spider (whence the Arachnida of our Entomology) her rope into

a web, and left her in perpetua to web, or as we now say, to weave.
Dante describing his visit to Inferno met her in the lowest woe,
and writes :—

> "O fond Arachne ! thee I only saw
> Half spider now in anguish crawling up
> The unfinished web thou weaved'st to thy bane."

From whatever causes our beautiful silk industry has met with
Arachne's fate, I feel sure that our new awakening to the necessity
of technical instruction and greater efficiency shall one day restore
our Arachne to her loom and real shape, and begin again to twist and
weave that charming thread of which some one has aptly said
that silk is to the fibres what gold is amongst metals, and the
diamond amongst jewels, and that we shall hope to see our country
by art creative power and thoughtful skill as lastingly famous as
were Moir and Mohair from the Moors, Cambric from Cambay,
Bokhara for its buckram, Renne for its cloth of Rayne, Cyprus for
its cypresse, Tucker-street, Bristol, for its tuck, the Saracens for
sarsenets, Calicut for its calicoes, Tartarium cloths from Tartary,
so skilfully woven that

> "No painter's brush could match them hanging in
> Broad bands of fine Tartarians."

so Chaucer wrote Fostat for its fustians, Arras for its arras,
Nankin for its nankeen, Gaza for its gauze, Cordova for its cord-
wain, Baiae for its baize, Friesland for its frieze, Jean for its jean,
Damietta for its dimity, Tarsus for its tabriz, Droghcda for its
druggets, Old Worsted (in Norfolk) for its worsted, Kersey for its
kerseymere, Linsey for its linsey-wolsley, Guninghamp for its ging-
ham, Avignon for its papal cloth or poplin, five hundred years ago,
Masul for its muslin, Damascus for its damask. Of ancient places,
many still famous, as Cashmere for its shawls, and formerly for its
kerseymere, Coptic Akhmim for its tunics and cloths, Dacca for its
floss and finest of muslins.

Let us augur that this beautiful but almost lost industry may be
soon regained, and that the £11,000,000 sterling we have been for
the last 25 years paying the foreigner for our manufactured silks,
shall be paid to a future and successful British industry, and that
if the ladies will only come to our aid and prefer British art in
silk, we shall be able to say that "they shall walk in silk attire
and siller hae to spare."

LIST OF THE LANTERN SLIDES.

1. Cotton, showing its flat and twisted fibre of cellulose.
2. Linen Fibres, the elementary fibres of flax (Linum usitatissi mum) are cells of pellucid membranes joined end to end with can-like joints, between which form thickened vessels with fibrous matter; along with these are seen the woody fibre or tissue, consisting of "elongated cells or tubes, with tapering extremities which overlap each other, and by their union longitudinally form the fibres called hemp and flax."
3. Typical Woollen Fibre.
4. Spider Silk.
5. Silk of Saturnia carpini.
6. Silk Bave and Brin in contact and separated, also showing sections.
7. Bave of Bengal silk, showing loops as scriposited, demonstrating that the fibre is regular.
8. Rough places in the silk fibre formerly supposed to be inherent, shown under the microscope to be merely unresolved loops in rabble, caused by imperfect reeling.
9. Silk Fibres, Bave and Brin shown to be perfectly smooth, although somewhat irregular in thickness.
10. Larva of Bombyx Mori feeding on the mulberry leaf.
11. Bengal cocoons of Bombyx fortunatus.
12. Avenues of twigs upon which the larva of Bombyx Mori form their cocoons in Italy.
13. Interior of magnaneric or cocoon sorting room.
14. Oven or chamber in which the chrysalides are killed to prevent them escaping from the cocoons.
15. A Native Indian Wheel for silk reeling.
16. Improved European Machinery for cocoon-reeling with the Tavelette Keller, and girl reeling.
17. Silk Moths of the Bombyx Mori.
18. Larva of Antheræa mylitta, or Tussur Silkworm.
19. Cocoons of the Tussur Silkworm.
20. Cocoons of the Tussur Silkworm attached to branches.
21. Indian Tussur Silk Bave, showing fibrillæ and sections.
22. Indian Tussur Silk Brin, showing fibrillæ separated.

23. Indian Tussur Silk Bave, showing foldings of the fibres as laid by the silkworm.
24. Indian Tussur Silk Moth, Antheraea mylitta, male.
25. Indian Tussur Silk Moth, Antheraea mylitta, female.
26. Cocoons of Philosamia ricini, or Eria Silkworm.
27. Fibres from cocoons of ditto.
28. Moth of Philosamia ricini, or Eria Silk moth.
29. Anthræopsis assama, or Muga Cocoons.
30. Moth of Anthræopsis assama, or Muga Moth, male.
31. Cocoon of Rinacula zulica.
32. Moth of ditto.
33. Cocoon of Actias selene.
34. Moth of Actias selene, male.
35. Moth of Actias selene, female.
36. Moth of Actias leto, male.
37. Moth of Actias leto, female.
38. Cocoon of the Atlas Moth, Attacus Atlas.
39. Attacus Atlas Moth, female, the largest moth known.
40. Moths of Bombyx Mori, for size comparison with Attacus Atlas.
41. Scales from the moth's wings magnified.
42. A single scale from the moth's wings highly magnified.
43. Silk Weaving at Agra.
44. Jewelled and Spangled Cloth Embroiderers at work, Delhi.
45. Old Italian Embroidery.
46. Silk Embroidered Chasuble.
47. Old Italian Embroidery.

LIST OF SILK-PRODUCING LEPIDOPTERA,

BROUGHT UP

TO THE PRESENT DATE.

———

Those marked thus *a* are in my collection in the imago state. Of those marked *aa* I have both imago and cocoon. Of those marked thus † I have the larva, cocoon, and imago. Of those marked thus †† I have the imago, larva, and chrysalis. Of those marked thus ‡ I have the cocoon only. Those not marked I do not possess in any stage, and I shall be glad to receive larvæ and mothes but more particularly the cocoons of all or any of these species to assist me in the investigation of the physical properties and structure of the silk fibres of the Heterocera, a work in which I have for some time been engaged.

———

CLASSIFICATION.

ANIMALIA.

SERIES 2..	Metazoa (Huxley)
SUB-SERIES 2.................................	Deuterostomata *a Schizocæla*
SUB-KINGDOM 5.....................................	Anthropoda *Articulata*
CLASS 4...	Insecta
ORDER 7.............	Lepidoptera *Glossata*
SUB-ORDER...............	Heterocera (Moths)
Group..	Bombycina

FAMILY.	GENUS AND SPECIES.	LOCALITY.	
1 Sesiidae	Trochilium apiforme	Europe	†
2 Hetergynidae	Hetergynis penella	,,	aa
,,	,, paradoxa	,,	a
3 Zygaenidae	Ino chloros	,,	a
,,	,, globulariae	Britain	a
,,	,, statices	,,	a
,,	,, geryon	,,	a
,,	Zygaena pilosellae	,,	a
,,	,, exulans	,,	a
,,	,, meliloti	,,	a
,,	,, trifolii	,,	a
,,	,, lonicerae	,,	
,,	,, filipendulae	,,	aa
,,	,, carniolica	Europe	aa
,,	,, fausta	,,	aa
4 Nycteolidae	Sarrothripa undulana	,,	aa
,,	Erias vernana	,,	a
,,	,, clorana	Britain	a
,,	Hylophila prasinana	,,	aa
,,	,, bicolorana	,,	a
5 Lithosidae	Nola cucullatella	,,	a
,,	,, togatulalis	,,	††
,,	,, strigula	,,	a
,,	,, confusalis	,,	a
,,	,, albula	,,	a
,,	,, centonalis	,,	a
,,	Nudaria murina	,,	aa
,,	Setina irrorella	,,	a
,,	Lithosia mesomella	,,	a
,,	,, muscerda	,,	a
,,	,, griseola	,,	a
,,	,, stramineola	,,	a
,,	,, deplana	,,	a
,,	,, lurideola	,,	a
,,	,, complana	,,	a
,,	,, caniola	,,	a
,,	,, pygmaeola	,,	a
,,	,, sororcula	,,	aa
,,	,, quadra	,,	a
,,	Gnophria rubricollis	,,	a
6 Arctiidae	Emydia striata	,,	a
,,	,, cribrum	,,	a

21

FAMILY.	GENUS AND SPECIES.	LOCALITY.
6 Arctiidae	Nemeophila russula	Jersey *a*
,,	,, plantaginis	,, *a*
,,	Callimorpha dominula	Britain *a*
,,	,, hera	,, *a*
,,	Pleretes matronula	Europe
,,	Arctia caja	Jersey *aa*
,,	,, flavia	Europe
,,	,, villica	Jersey *a*
,,	,, purpurata	Europe †
,,	,, hebe	,, .††
,,	,, aulica	,, ††
,,	,, casta	,,
,,	Ocnogyna parasita	,, †
,,	Spilosoma fulginosa	Jersey *a*
,,	,, sordida	Alps
,,	,, mendica	Jersey *a*
,,	,, lubricipeda	,, *a*
,,	,. zatima	Europe *a*
,,	,, menthrasti	,, *a*
,,	,, urticae	,, *a*
7 Hepialidae	Hepialus virescens	New Zealand
8 Cossidae	Cossus cossus	Europe
,,	,, terebra	.,
9 Cochliopodae	Heterogenea limacodes	Britain *a*
,,	,, asella	,, *a*
10 Psychidae	Psyche cocoons	Senegal
,,	,, cocoons found in the tea plant	India
,,	,, cocoons	Natal
,,	,, cocoons	,,
,,	,, cocoons	Dominica
,,	,, unicolor	Europe †
,,	,, villosella	,, ††
,,	,, ecksteini	,, ††
,,	,, zelleri	,, ††
,,	,, plumifera	,, ††
,,	,, hirsustella	,, ††
,,	,, cocoons	Trinidad
,,	Thyridopteryx ephemerae-formi	North America
,,	Euncta japonica	India
,,	Epichnopteryx helicinella	Europe

FAMILY.	GENUS AND SPECIES.	LOCALITY.
11 Liparidae	Orgyia gonostigma	Britain
,,	,, antiqua	,,
,,	,, ericae	Europe †
,,	,, josephina	Algeria a
,,	,, leucostigma	North America
,,	Dasychira confusa	Amur
,,	,, selenitica	Europe
,,	,, pudibunda	Britain
,,	,, fascelina	,,
,,	,, abietis	Europe
,,	Laelia coenosa	Britain
,,	Laria L nigrum	Europe
,,	Leucoma salicis	Britain
,,	Porthesia chrysorrhœa	,, a
,,	,, auriflua	,, a
,,	Psilura monacha	,, a
,,	Ocneria dispar	,, a
,,	Darala ocellata	Indo-Australasia a
,,	Murlida mutans	,, a
,,	,, citrina	,, a
,,	,, undata	,, a
,,	Calepteryx collesi	,, a
12 Bombycidae	Bombyx mori	Italy, green cocoons †
,,	,,	Italy, yellow cocoons aa
,,	,,	Italy, white cocoons aa
,,	,,	Japan, white cocoons ‡
,,	,,	China, white cocoons ‡
,,	,,	French, white & yellow cocoons ‡
,,	,,	Broussa, white & yellow cocoons ‡
,,	,,	Cyprus, large white cocoons ‡
,,	,	Cashmere cocoons ‡
,,	,,	Burma cocoons ‡
,,	,,	Ceylon cocoons ‡

FAMILY.	GENUS AND SPECIES.	LOCALITY.
12 Bombycidae	Bombyx mori	Kashmircocoons ‡
,,	,,	Saharumpur ,, ‡
,,	,,	Cawnpur ,, ‡
,,	,,	Lahore ,, ‡
,,	,,	Persia ,, ‡
,,	,, fortunatis	Bengal *aa*
,,	,, crœsi	,, *aa*
,,	,, textor	India *aa*
,,	,, meridionalis	,, *aa*
,,	,, arracanensis	Arracan and Burmah *aa*
,,	,, sinensis	India *aa*
,,	,, rhadama	Madagascar *aa*
,,	,, borocera	North America
,,	,, crataegi	Britain *aa*
,,	,, populi	,, *aa*
,,	,, francomcia	Europe *aa*
,,	,. alpicola	,, *a*
,,	,, castrensis	Britain †
,,	,, neustria	,, †
,,	,, americana	North America *a*
,,	,, innocens	,, ,, *a*
,,	,, henckei	Rossmer *a*
,,	,, lanestris	Britain †
,,	,, catax	Europe †
,,	,, rimicola	,, †
,,	,, trifolii	Britain †
,,	,, abserrula	Spain †
,,	,, quercus	Britain *a*
,,	,, rubi	Britain *a*
,,	,, ursula	Africa *a*
,,	,, ilicis	Europe
,,	,, neogene	,,
,,	,, loti	,,
,,	,, vandalicia	,,
,,	,, fasciatella	,,
,,	,, affinis	Chota Nagpur
,,	,, annulipes	Africa
,,	,, bauhinia	,,
,,	, diego	,,
,,	,, fleurioti	,,
,,	,, panda	,,

FAMILY.	GENUS AND SPECIES.	LOCALITY.
12 Bombycidae	Chondrostega pastrana	Europe
,,	Crateronyx dumi	,, a
,,	Lasiocampa potatoria	Britain †
,,	,, albomaculata	Amoor a
,,	,, laeta	,, a
,,	,, pruni	,, aa
,,	,, sumatrensis	Indo-Australasia a
,,	,, quercifolia	Britain †
,,	,, populifolia	Central Europe †
,,	,, tremulifolia	Europe a
,,	,, ilicifolia	Britain a
,,	,, new species	Indo-Australasia a
,,	,, suberifolia	France & Spain a
,,	,, pardale	Indo-Australasia a
,,	,, lunigera	South Germany a
,,	,, ablobulina	Saxony a
,,	,, pini	Central Europe a
,,	,, bufo	Europe & Siberia †
,,	,, otus	,, ,, †
,,	,, dieckmanni	Siberia a
,,	,, lineosa	Europe
,,	,, femorata	,,
,,	Megasoma repanda	,, a
,,	Theophila huttoni	India aa
,,	,, mandarina	China
,,	,, sherwilli	S. E. Himalayas
,,	,, bengalensis	Lower Bengal
,,	Ocinara religiosa	Assam & Cachar
,,	,, comma	Doon Mussooric
,,	,, livida	Mussooric
,,	Trabala vishnu	India
,,	,, villosipes	,,
,,	Trilochana varians	Calcutta
,,	Borocera postica	Madagascar
,,	Eucheira socialis	Mexico
,	Demas coryla	Britain a
,,	Eriogaster proximo	South America a
,,	,, hirtina	,, ,, a
,,	Hylesia canitiae	Honduras †
,,	Titya undulosa	South America a
,,	Macromphalia chilensis	,, ,, a
,,	Boroceros postica	South Africa †

FAMILY.	GENUS AND SPECIES.	LOCALITY.
12 Bombycidae	Lebedaa pityocampacramcis	Africa *a*
,,	Gangerarides rosea	Indo-Australasia *a*
,,	Drymonia senatoria	North America *a*
,,	,, alba	,, ,, *a*
,,	Nemeresa trimacula	South America *a*
,,	Hydrias nacens	,, ,, *a*
,,	Gonometa postica	Africa *a*
,,	Palustra burmeisteri	South America *aa*
13 Endromidae	Endromis versicolora	,, ,, †
14 Saturnidae	Attacus atlas	India †
,,	,, cynthia	,, †
,,	,, ricini	,, *aa*
,,	,, insularis	,, *a*
,,	,, bolivar	Trinidad
,,	,, hesperus	Guiana *a*
,,	,, aurota	South America *aa*
,,	,, jacobae	,, ,, *a*
,,	,, maurus	,, ,, *a*
,,	,, orizaba	Mexico *a*
,,	,, speculum	South America *aa*
,,	,, sp.	Demerara (G. T. Hawtayne,Esq.)*aa*
,,	,, arethusa	South America *a*
,,	,, canningi	Hindostan .
,,	,, edwardsia	Sikkim
,,	,, guerini	India
,,	,, lunula	Ceylon
,,	,, obscurus	Cachar
,,	,, splendidus	Africa
,,	,, jorulla	Mexico
,,	,, crameri	Amboina
,,	,, andromeda	Valdivia
,,	,, cinerascens	,,
,,	,, larquinii	Lozou
,,	,, vesta	Hindostan
,,	,, walkeri	Ning-po
,,	,, chapata	Mexico
,,	,, betis	Brazil
,,	,, aricia	Santa Fe de Bogota
,,	,, ethra	Caraccas
,,	,, zacatica	Bogota
,,	,, vacuna	Ashantee

FAMILY.	GENUS AND SPECIES.	LOCALITY.
14 Saturnidae	Attacus mythimna	Zululand
,,	,, baumhiria	Senegal
,,	,, irius	East Indies
,,	,, saturnus	,, ,,
,,	,, sylhetica	Sylhet
,,	,, splendens	Bogata
,,	,, laventera	Mexico
,,	,, gelleta	,,
,,	Platysamia cecropia	North America aa
,,	,, columbia	Columbia a
,,	,, ceonothi	North America a
,,	,, calleta	South America a
,,	,, gloveri	North America a
,,	Callosamia angulifera	,, ,, aa
,,	Samia promethea	,, ,, aa
,,	Telea polyphemus	,, ,, aa
,,	Bunaea alcinoe	Africa a
,,	,, caffaria	South Africa a
,,	,, tyrrhena	Zululand
,,	,, alinda	Sierra Leone
,,	,, phaedusa	West Africa
,,	,, nictitans	Tropical Africa
,,	,, angusana	Port Natal
,,	,, alopia	South Africa
,,	,, farda	Port Natal
,,	,, acetes	Cape of Palms
,,	,, mopsa	W. Africa
,,	,, dorcas	,,
,,	,, epimethea	Ashanti
,,	,, magniferae	Madagascar
,,	,, zambesina	Zambesi region
,,	Copaxa decresens	South America a
,,	,, canella	Brazil
,,	,, satellitiae	Bogota
,,	,, expandens	Venezuela
,,	,, plenkeri	Mexico
,,	Syntherata janetta	Indo-Austra-lasia aa
,,	Sagana sapatoza	Santa Fe de Bogota
,,	,, punctigera	,, ,,
,,	Thyelia nyctelops	Caffraria
,,	,, bunaea	

FAMILY.	GENUS AND SPECIES.	LOCALITY.	
14 Saturnidae	Antheraea arata	Ashanti	
,,	,, cytherea	W. & S. Africa	
,,	,, hersilia	Congo	
,,	,, tyrrhea	S. Africa	
,,	,, dione	Africa	
,,	,, suraka	Madagascar	
,:	,, larissa	Java	
,,	,, jana	Java	
,,	,, astrocephala	Australia	
,,	,, rumphi	Amboina	
,,	,, semper	Luzon	
,,	,, purpurascens	N. Australia	
,,	,, disjuncta	Moreton Bay	
,,	,, pristina	New Guinea	
,,	,, walbergi	S. Africa	
,,	,, helena	Tasmania	
,,	,, andamani	S. Andamans	
,,	,, confuci	Shanghai	
,,	,, mezankooria	Assam	
,,	,, nebulosa	Singbhoom	
,,	,, perrotteti	Pondicherry	
,,	,, assama	Assam	aa
,,	,, mylitta	India	aa
,,	,, frithii	Sikhim	a
,,	,, helferi	,,	a
,,	,, menippe	Africa	a
,,	,, roylei	India	aa
,,	,, pernyi	China	aa
,,	,, feltoni	,,	a
,,	,, belina	Indo-Australasia a	
,,	,, yama-mai	Japan	aa
,,	,, eucalypti	Indo-Australasia a	
,,	,, simplex	Africa	a
,,	Caligula simla	India	aa
,,	,, japonica	Japan	
,,	,, cachara	Cachar	
,,	,, thibeta	Thibet	
,,	Gyanisia isis	Africa	a
,,	Neoris huttoni	N.W.Himalayas a	
,,	,, shadulla	Yarkund	
,,	,, stolughana	Ladak India	
,,	Loepa katinka	India	†

FAMILY.	GENUS AND SPECIES.		LOCALITY.	
14 Saturnidae	Leopa miranda		India	aa
,,	,, sikkima		Sylhet	
,,	,, sivalica		Mussooro	
,,	Rhodia newera		Sikkim	†
,,	Hyperchiria io		NorthAmerica	aa
,,	,, pamina		South America	a
,,	,, metzlii		,, ,,	a
,,	,, liberia		,, ,,	a
,,	,, orodes		,, ,,	a
,,	,, scapularis		,, ,,	a
,,	,, coresus		,, ,,	a
,,	,, viridescens		,, ,,	a
,,	,, euryopa		,, ,,	a
,,	,, illustris		,, ,,	a
,,	,, complicata		,, ,,	a
,,	,, salmonia		,, ,,	a
,,	,, beskei		,, ,,	a
,,	,, nyctimena		,, ,,	a
,,	,, aspera		,, ,,	a
,,	,, myops		,, ,,	
,,	,, braziliensis		Brazil	
,,	,, erythrina		Valdivia	
,,	,, larra		Rio Janeiro	
,,	,, junonia		Bagota	
,,	,, convergens		Rio Janeiro	
,,	,, beckeri		S. America	
,,	,, oblonga		Santa Fe de Bogota	
,,	,, jucunda		Surinum	
,,	,, griseoflava		Chili	
,,	,, tridens		Brazil	
,,	,, basalis		Venezuela	
,,	,, melanops		Brazil	
,,	,, incarnata		Bogota	
,,	, cruenta		Brazil	
,,	,, approximata		Bogota	
,,	,, memusea		Brazil	
,,	,, acutissima		Mexico	
,,	,, cinerea		Brazil	
,,	,, megalops		Bogota	
,,	,, pyrrhomelas		Santa Fe de Bogota	
,,	,, combusta			
,,	,, inornata		Brazil	

FAMILY.	GENUS AND SPECIES.		LOCALITY.
14 Saturnidae	Hyperchiria	luteata	
"	"	janeira	Rio Janeiro
"	"	hebe	Oajaca Mexico
"	"	abas	Surinum
"	"	iris	Oajara Mexico
"	"	continua	Mexico
"	"	arminia	Surinum
"	"	inficita	Ega
"	"	incisa	Brazil
"	"	saturniata	Bogota
"	"	nausica	Brazil
"	"	erythrops	Coquimbo
"	"	metapyrrha	Rio Janeira
"	"	approximans	" "
"	"	subcana	" "
"	"	schaussi	N. America
"	"	varia	" "
"	"	ruberescens	Venezuela
"	"	saturata	Mexico
"	"	janus	Surinum
"	"	fusca	S. America
"	"	bilinea	Para, Brazil
"	"	armida	Surinum
"	"	metea	Bolivia
"	Brahmaea lunulata ledereri		Europe a
"	"	certhia	Sylhet
"	"	lucina	Sierra Leone
"	"	undulata	Ning-po
"	Saturnia boisduvali		Europe a
"	"	atlantica	" ‡
"	"	pyri	South Europe a
"	"	schenkii	" " a
"	"	anna	Sikkim
"	"	carpini	
"	"	lindia	Sikkim
"	"	pyretorum	Siberia a
"	"	spini	S. E. Europe and
"			W. Asia a
"	"	grotei	Darjeeling
"	"	galbina	South America a
"	"	jankowskii	Europe a

FAMILY.	GENUS AND SPECIES.	LOCALITY.
14 Saturnidae	Saturnia diana	Europe and Siberia *aa*
,,	,, mendocina	North America *a*
,,	,, caecigena	Europe & Western
,,		Asia *aa*
,,	,, fuscicolor	Africa
,,	,, nenia	Congo
,,	,, said	Bagamayo
,,	,, cajana	Africa
,,	,, auricolor	,,
,,	,, wallengrenii	Caffraria
,,	,, usta	
,,	,, attacus	S. Africa
,,	,, hoegii	S. America
,,	,, lucina	Sierra Leone
,,	,, melvilla	Melville Island
,,	,, cidosa	Sikkim
,,	Actias isabella	Spain *a*
,,	,, artemis	Siberia *a*
,,	,, luna	North America *aa*
,,	,, selene	India *aa*
,,	,, ignescens	Andaman Isles
,,	,, leto	East Indies
,,	,, manas	Sylhet
,,	,, sinensis	North China
,,	,, dictynna	
,,	,, mimosa	S. Africa
,,	,, rosenbergii	Amboina
,,	Aglia tau	Central Europe &
,,		Northern Asia *a*
,,	Rinaca zuleika	Sylhet
,,	Salassa lola	,,
,,	Endaemonia semiramis	Honduras
,,	,, argus	Ashanti
,,	Perisomena semicaecia	Port Natal
,,	Giacla	
,,	Polysthana rubrescens	South America *aa*
,,	,, andromeda	,, ,, *a*
,,	Hemileuca maja	North America *a*
,,	,, nevadensis	,, ,, *a*
,,	,, pica	United States
,,	Pseudohazis venosa	Carraccas

31

FAMILY.	GENUS AND SPECIES.	LOCALITY.
14 Saturnidae	Pseudohazis eglanteria	California
,,	Aphelia apollinaris	Port Natal
,,	Rhescyhtis hyppodamia	Rio Janeiro
,,	,, armida	S. America
,,	,, hercules	Rio Janeiro
,,	,, aspasia	S. America
,,	,, sylla	Surinum
,,	,, meander	Brazil
,,	,, xanthopus	,,
,,	,, latifascia	
,,	,, kadenii	
,,	,, erythrina	South America a
,,	,, pandora	,, ,, a
,,	Eudelia rufescens	,, ,, a
,,	Urota sinope	Africa a
,,	Dysdaemonia glaucescens	Brazil
,,	,, boreas	West Indies
,,	Henucha smilax	Port Natal
,,	,, grimmia	S. Africa
,,	,, delegarguei	Port Natal
,,	Heliconisa impara	Spirito Sancto
,,	,, pagenstecheri	
,,	Micrattacus nanus	Rio Janeiro
,,	,, dissimilis	South America a
,,	Mimallo despecta	South America a
,,	,, amilio	Brazil
,,	,, cicinnus	Chili
,,	,, arthane	Conception
,,	,, plana	Brazil
,,	,, verago	Surinum
,,	,, saturata	Rio Janeiro
,,	,, trilunula	Brazil
,,	,, plagiata	,,
,,	Microgone agathylla	Congo
,,	Cyrtogone naenia	S. Africa
,	,, herilla	Africa a
,,	Molippa sabina	South America a
,,	Coloradia venata	,, ,, a
,,	Dirphia tarquinia	Cayenne aa
,,	,, vulpina	South America a
,,	,, cinnamonea	,, ,, a
,,	,, marginata	,, ,, a

FAMILY.	GENUS AND SPECIES.	LOCALITY.
14 Saturnidae	Dirphia glauca	South America a
,,	,, speciosa	,, ,, aa
,,	,, calchas	Rio Janeiro
,,	,, triangulum	,, ,,
,,	,, arasia	Surinum
,,	,, avia	,,
,,	,, polybia	,,
,,	,, ursina	Bolivia
,,	,, concolor	Venezuela
,,	,, varia	Bolivia
,,	,, eumenides	Surinum
,,	,, obtusa	Para
,,	,, rustica	Surinum
,,	,, nubila	,,
,,	,, agis	Brazil
,,	,, pulchricornis	Venezuela
,,	,, semirosea	Mexico
,,	,, somniculosa	Brazil
,,	,, litura	Santa Fe de Bogota
,,	,, hircia	Surinum
,,	,, amalia	,,
,,	,, plana	,,
,,	,, angulifera	Peru
,,	,, multicolor	Rio janeiro
,,	,, rosacordis	,, ,,
,,	,, satellitia	,, ,,
,,	,, purpurascens	,, ,,
,,	,, andulosa	,, ,,
,,	,, vagans	Brazil
,,	,, aconyta	Bengal, Coromandel
,,	,, canitia	Surinum
,,	,, phadima	,,
,,	,, metabus	,,
,,	,, obsoleta	,,
,,	,, rivulosa	,,
,,	,, cognata	Valdivia
,,	,, maginata	,,
,,	,, melanostigma	Brazil
,,	,, maginella	Bogota
,,	,, quadricolor	Ega
,,	,, pallida	Bogota

FAMILY.	GENUS AND SPECIES.	LOCALITY.
14 Saturnidae	Caloptera ocellata	Crete
,,	Ceratocampa regalia	South America *a*
,,	,, imperialis	North America *a*
,,	Rhodia newera	India
	DOUBTFUL SILK PRODUCERS IN THIS FAMILY.	
,,	Eacles or Ceratocampa	
,,	,, princeps	Brazil
,,	,, laocoon	Georgia
,,	,, ducalis	Rio Janeiro
,,	,, magnificas	Para
,,	,. pharonea	Brazil
,,	,, cacicus	Rio Janeiro
,,	,, principalis	Brazil
	SILK PRODUCERS.	
15 Drepanulidae	Drepana lacerta	Britain *aa*
,,	,, harpagula	,, *a*
,,	,, falcataria	,, *aa*
,,	., binaria	,, *a*
,,	,, cultraria	,, *a*
,,	Cilix glaucata	Europe *a*
,,	Cricula trifenestrata	India †
,,	Lonomoia albigutta	South America *a*
16 Notodontidae	Harpyia furcula	Britain ‡
,,	,, bifida	,, ‡
,,	,, vinula	,, ‡
,,	Stauropus fagi	,, *a*
,,	Notodonta tremula	,,
,,	,, dictaeoides	,, *a*
,,	,, ziczac	,, *a*
,,	,, tritophus	,, *aa*
,,	,, trepida	Europe *aa*
,,	,, torva	,, *aa*
,,	,, dromedarius	Britain *a*
,,	,, chaonia	,, *a*
,,	,, dodonea	,, *aa*
,,	,, bicolora	,, *a*
,,	Cophopteryx carmelita	,, *a*

34

FAMILY	GENUS AND SPECIES.	LOCALITY.	
16 Notodontidae	Cophopteryx camelina	Britain	a
„	„ cucullina	„	a
„	Uropus ulmi	„	††
„	Asteroscopus sphinx	„	a
„	Pterostoma palpina	„	a
„	„ nubeculosus	„	a
„	Gluphisia crenata	„	a
„	Ptliophora plumigera	„	a
„	Pygaera anastomosis	„	††
„	„ curtula	„	
„	„ anachoreta	„	
„	„ pygra	„	
„	Heterocampa subrotata	North America	a
„	„ quadrata	South America	a
„	„ amazonic	„ „	a
„	Edema albifrons	North America	a
„	Cnethocampa processionea	South & Central Europe	a
„	„ pityocampa	North Germany	a
„	„ pinivora	Europe	aa
„	„ herculeana	„	a
17 Cymatophoridae	Thyatira batis	Europe	†
„	Asphalia ridens	„	
18 Noctuæ	Loma orion	Britain	
„	Dipthera ludifica	Europe	aa
„	Acronycta auricoma	Britain	a
„	„ menyanthidis	„	a
„	Simyra venosa	„	†
„	Diloba caeruleocephala	Britain	
„	Arsilonche albovenosa	„	
„	Clidia geographica	„	†
„	Polia canescens	„	†
„	Panthea coenobita	„	†
„	Memestra persicariae	Saxony	
„	Valeria oleagina	Europe	†
„	Hadnea porphyrea	„	†
„	Eriopus latreilli	„	aa
„	Trachea atriplicis	„	a
„	Euplexia lucipara	„	aa
„	Dicycla oo	„	††
„	Cucullia verbasci	„	
„	„ scrophulariae	Britain	a

FAMILY.	GENUS AND SPECIES.	LOCALITY.	
18 Noctuœ	Cucullia lactucae	Europe	a
,,	,, chamomillae	,,	
,,	,, tanecti	,,	
,.	,, argentea	,,	†
,,	Eurhipia adulatrix	,,	†
,,	Plusia triplasia	,,	†
,,	,, moneta	,,	aa
,,	,, festucae	,,	†
,,	Aedia funesta	,,	†
,,	Thalpohares dardouini	,,	†
,,	,, rosea	,,	
,,	Erastria venustula	,,	aa
,,	Catocala sponsa	Britain	a
,,	,, promissa	Europe	†
,,	Spintherops spectrum	,,	†
19 Geometrae	Odontopera bidentata	,,	aa
,,	Eugonia autumnaria	,,	aa
,,	Eugonia fuscantaria	,,	a
,,	Epione Apiciaria	,,	††
,,	Pericallia syringariae	Europe	††
,,	Rumia crataegata	Britain	
,,	Acidalia luteola	,,	‡
,,	Numeria pulverari	,,	‡
,,	Cidaria sagittata	,	‡
,,	Eucosmia certata	,,	‡
,,	Eupithecia togata	,,	‡
20 Galleriae	Galleria mellonella	,,	
21 Pyralidae	Eurycreon turbidalis	Hungary	‡
22 Tortricina	Tortrix viridana	,,	a
23 Talaeporidae	Talaeporia politella	Europe	aa
,,	,, pseudobombycella	,,	†
,,	Solenobia clathrella	,,	aa
,,	,, pineti	,,	aa
,,	,, triquetrella	,,	†
24 Lypusidae	Psilothrix dardoinella	,,	aa
25 Tineidae	Tinea pellionella	,,	‡
26 Hyponomeustidae	Hyponomeuta evonymellus	Britain	a
	,, padi	,,	
,,	,, plumbellus	,,	a
,,	,, padellus	,,	a
27 Pentellidae	Cerostoma denitella	,,	
,,	Harpipteryx harpella	,,	
,,	Plutella xylostella	,,	

LIST OF SILK AND OTHER EXHIBITS.

1. Entomological case of arranged moths, larvæ, cocoons, and silk : *a.* Larva of Tussur silk moth, Antheraea mylitta ; *b.* Male and female moths of Antheraea mylitta ; *c.* Cocoons, whole, pierced and showing chrysalides ; *d.* Tussur raw silk, native reeled at Fatwah ; *e.* ditto reeled by the Italian method by Messrs. Louis Payen and Co., at Behrampore ; *f.* Male and female moths of the Chinese Tussur silkworm, Antheraea pernyi ; *g.* Cocoons of Tussur silkworm, Antheraea pernyi ; ditto in oak leaves ; *h.* Chinese reeled silk of Tussur silkworm, Antheraea pernyi ; *i.* Larva of Bombyx mori ; *j.* Male and female moths of Bombyx mori ; *k.* Cocoons of the mulberry-fed silkworm, Bombyx mori, Italian, white ; *l.* ditto, yellow ; *m* Hot-weather cocoons, Bombyx cræsi, and cold-weather cocoons Bombyx fortunatus of Bengal.

2. Large diagram of larvæ, moths, and cocoons, natural size and enlarged, and raw-silk : *a.* Tussur silkworm or larva ; *b.* Tussur moth, Antheraea mylitta ; *c.* Tussur cocoon showing pedicular attachment ; *d.* Chinese Tussur moth, Antheraea pernyi; *e.* Chinese Tussur cocoon ; *f.* Mulberry fed silkworm of commerce, Bombyx mori ; *g.* Bombyx mori moth.

3. Large drawing of native Arrah women reeling and winding Tussur silk by their native methods.

4. Large drawing of the European method of cocoon reeling with the Italian Tavelette Keller.

5. Diagram of the microscopic appearance of tussur silk fibre, longitudinally and in section. Ditto, ditto showing separated fibrillæ.

6. Ditto, ditto, of the ordinary silk of commerce, Bombyx Mori, longitudinally and in section.

7. Five bottles containing silkworms in spirit : *a.* Larva of Bombyx mori ; *b.* Larva of the Tussur moth, Antheraea mylitta ; *c.* Larva of Cricula trifenestrata ; *d.* Larva of the eri moth, Attacus ricini ; *e.* Larva of the Muga moth, Antheraea assama.

8. Grapelike clusters of the cocoons of Cricula trifenestrata, from Ranchi, Chutia Nagpur, Bengal.

9. Three hand-reels used by the natives for reeling Tussur cocoons. One hand-reel used by natives for reeling ordinary cocoons.

10. The Italian Tavelette Keller.

11. Map of India.

12. Convoluted flat tape of tin-plate showing scintillations of Tussur silk.

13. Cylindrical rod of the same material, showing equal diffusion of light.

14. Santal bow and arrow.

15. Hortus siccus of the tussur food-plants, with tussur cocoons attached to the branches, collected by the Rev. A. Campbell in the Santal Jungles in Gobinpore, Manbhum : a. Terminalia tomentosa ; b. Terminalia chebula ; c. Terminalia arunja ; d. Terminalia bellerica ; e. Shorea robusta ; f. Zizyphus jujuba ; g. Lagestroemia parviflora ; h. Careya arborea ; i. Diospyros tomentosa ; j. Alstonia scholaris.

16. Tussur cocoons from Gaya.

17. Samples of Tussur raw silk reeled by Messrs. Louis Payen and Co., at Behrempore.

18. Woodcutter's knife-belt with jingle-bells of tussur cocoons.

19. Tussur silk spun into yarn for weaving into sealcloth ; three sizes, dyed and undyed, manufactured and lent by Messrs. J. Bradwell & Co., silk spinners and merchants, Congleton.

20. Seal plush.

21. Dinner-table centre on satin, worked in Indian Tussur silk, designed by the late John Sedding.

22. Tussur silk reeled at Fatwah.

23. Four specimens of 1 metre each, of brocades, the ornamenta parts of Indian Tussur silk, woven in patterns of Louis XVth and Louis XVIth styles ; manufactured by Messrs. Lamy & Giraud, Lyons.

24. Small example of gold-coloured damask, manufactured by Messrs. Perkin & Sons, 27, Curtain Road, London, E.C.

25. Example of cream-coloured figured damask, Indian Tussur silk, manufactured by Messrs. M. Perkin & Sons, 27, Curtain Road, London, E.C.

26. Three samples of Indian Tussur silk, figured damask, woven in Leek, in 1879.

27. Two specimens of French-made open work dress damasks, Indian Tussur silk.

28. Tussur Silk Chenille in various colours, and cut in a variety of patterns.

29. A series of Tussur Chenilles, chenille fringes, trimmings, tassels, toy monkeys, and swans, deep fringes of wool and tussur mixed, and pom-poms manufactured and lent by Messrs. W. and G. Kessler, Berlin.

30. Series of Satin Damask and Brocade of Indian Tussur silk, manufactured by Messrs. Devaux and Bachelard, of Lyons, 1 metre each.

31. Six samples of Indian native woven bleached Tussur silk, and three examples of Indian native woven unbleached Tussur silk for dresses ; printed by Thomas Wardle, Leek.

32. Indian Tussur silk tram, dyed to the colour of 14 precious stones, by G. C. Wardle, Leek.

33. A bundle of three threads organzine of Indian Tussur silk, manufactured by Messrs. J. and T. Brocklehurst and Sons, Macclesfield.

34. Lace, manufactured by Messrs. J. Kirkbride and Co., St. Mary's Gate, Nottingham ; the pattern part is made of Indian Tussur silk, thrown by Messrs. J. and T. Brocklehurst and Sons, Macclesfield.

35. Cream coloured bleached opera fichu, of Indian Tussur silk, lent by Mr. Maas, of Berlin ; also pocket handkerchief and pieces of Tussur silk for rain and dust cloaks, lent by Mr. Maas.

36. One 10-yard piece of native woven Tussur silk, " Dhoti " for men's wear, woven at Giridi, Bengal, undyed, with red border.

37. Indian Tussur raw silk, reeled by Messrs. Louis Payen and Co., at Behrempore.

38. Three bottles of powdered Tussur silk, for surface coating ; a new French utilization.

39. Native spun and woven undyed cloth, made of the silk of the Atlas worm, Attacus atlas, Nepal Terai ; red plaid on undyed ground.

40. Sozni, or Bedcover, embroidered at Peshawur.

41. Doopata (turban) of cotton, embroidered with undyed Muga silk. Worked at Dacca, Bengal.

42. Embroidered silk on cotton phulchari. Made and worn in Amristar, Punjaub.

43. Embroidered border on satin, from Cutch, Bombay.

44. Berlin Tussur Silk Shawl.

45. Bi-coloured net. A marriage robe, most curiously dyed red on one side and green on the other. Ulwar.

46. Leek re-production of the effect of dyeing a different colour on each side as in the Ulwar sample. By Bernard Wardle.

47. Very interesting and ancient Indo-Persian figure weaving.

48. Trimmings for ladies' hats, manufactured from Tussur silk.

49. Tussur silk cocoons attached to branches.

50. Cocoons of Anthraeaopsis Assama or Muga silkworm of Assam.

51. Raw silk   ditto   ditto   ditto

52. Italian cocoons of the Bombyx Mori from Bozzoli.

53. Bengal cocoons of Bombyx Fortunatus or Desi caterpillar, November bund.

54. Case containing samples of the improved reeling of Bengal silk.

55. Native woven Tussur silk cloth cleaned.

56 { Silken dwelling place of colony of Cnethocampa.
      Processionea or processional caterpillar. From the Riviera.

57. Turban, Bandana or tie and dye work, from Jeypore.

58. Ditto   ditto   ditto   Ulwar.

59. Gujrati Bandhana or tie and dye work in spots and borders of four colours, partly tied, from Jeypore.

60. Satin handkerchief, tie and dye work, from Jeypore.

61. Elaborate specimen of tie and dye or Bandhana work, Ulwar.

62. Illustration of the Great Buddhist Temple at Buddha-Gaya near Gaya, Bengal.

63. Ancient Coptic silk fabric IV century. From Echmine, Upper Egypt. Part of the clothing of rich Christian men and women of the time.

64. Modern reproduction of ditto   ditto   ditto.

65. Fragment of Silk found by Mr. Flinders Petrie, in a Christian coptic grave of the 6th century, A.D., at Fayum, Egypt, 1891.

66. Modern reproduction of ditto   ditto   ditto.

67. Pattern of the Silk Brocade with silver May blossoms, which has been lately woven in Spitalfields, by Messrs. Warner and Sons, for the bridal dress of Princess May of Teck.

68. Two designs found on the robe of St. Cuthbert, in his tomb in Durham Cathedral, date abont 1200, printed by Thomas Wardle.

69. Samples of modern silks woven at Spitalfields, by Messrs. Warner and Sons.

70. Samples of modern silks woven at Leek by Mr. G. H. Bermingham.

71. Cocoons of Coccidae, Eriopetis festucae, Signorctia luzula, Britain.

72. Case containing moths and cocoons of galleria cerella formed by larva in the honeycomb of the beehive.

73. Case containing moths and cocoons of Bombyx rhadama also sac or cocoon bag containing cocoons.